I0804222

Stem Cells to the Rescue

Helen M. Blau

Margery J. Fain

Cold Spring Harbor Laboratory Press

Stem Cells to the Rescue

All rights reserved
Text © 2024 Helen M. Blau; Illustrations © 2024 Margery J. Fain
Compilation © 2024 Cold Spring Harbor Laboratory Press, Cold Spring Harbor, New York
Printed in the United States of America

Publisher	John Inglis
Project Manager	Barbara Acosta
Editorial Assistant	Danett Gil
Permissions Coordinator	Carol Brown
Production Editor	Kathleen Bubbeo
Production Manager	Denise Weiss

Front cover image: Designed and illustrated by Margery J. Fain

Library of Congress Cataloging-in-Publication Data

Names: Blau, Helen M., author. | Fain, Margery J., illustrator.
Title: Stem cells to the rescue / Helen M. Blau and Margery J. Fain.
Description: Cold Spring Harbor, New York : Cold Spring Harbor Laboratory Press, [2024] | Includes bibliographical references and index. | Audience: Ages 7-11 | Audience: Grades 4-6 | Summary: "This book, intended for children ages 7 to 11, shows how stem cells have the potential to develop into many different types of cells and serve as a repair system in the body"— Provided by publisher.
Identifiers: LCCN 2024014864 (print) | LCCN 2024014865 (ebook) | ISBN 9781621825289 (paperback) | ISBN 9781621825296 (epub)
Subjects: LCSH: Stem cells—Juvenile literature. | Cellular therapy—Juvenile literature.
Classification: LCC QH588.S83 B58 2024 (print) | LCC QH588.S83 (ebook) | DDC 612.6/4018—dc23/eng/20240620
LC record available at https://lccn.loc.gov/2024014864
LC ebook record available at https://lccn.loc.gov/2024014865

10 9 8 7 6 5 4 3 2 1

All World Wide Web addresses are accurate to the best of our knowledge at the time of printing.

Authorization to photocopy items for internal or personal use, or the internal or personal use of specific clients, is granted by Cold Spring Harbor Laboratory Press, provided that the appropriate fee is paid directly to the Copyright Clearance Center (CCC). Write or call CCC at 222 Rosewood Drive, Danvers, MA 01923 (978-750-8400) for information about fees and regulations. Prior to photocopying items for educational classroom use, contact CCC at the above address. Additional information on CCC can be obtained at CCC Online at www.copyright.com.

For a complete catalog of all Cold Spring Harbor Laboratory Press publications, visit our website at www.cshlpress.org.

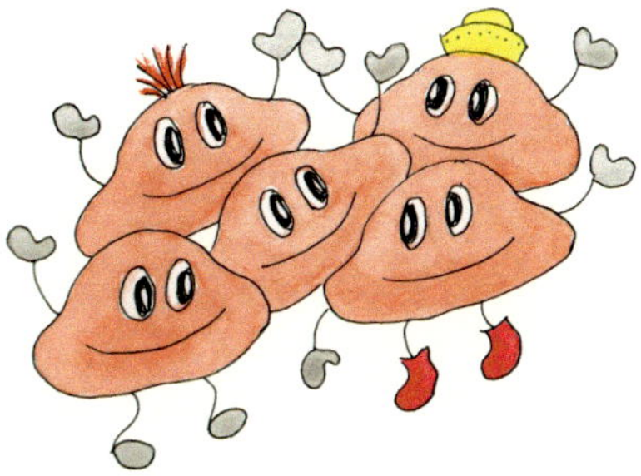

We dedicate this book to our children and grandchildren who inspire us every day with their curiosity and love of life.

The heroes of this book are stem cells. We chose frogs to tell the story, because many years ago frog cells showed scientists what stem cells can do.

Meet the Stem Cells

Meet the Frogs

Your body's made of special parts,
You need them all—for math, for arts,
Each one must do its job just right,
To play by day and sleep by night.

Your brain to think, your legs to walk,
Your arms to climb, and mouth to talk.
To skate, to read, to taste ice cream,
Hold yoga poses, and to dream.

You skateboard, playing violin.
Kapow! You hit the pickleball in.
You love to paint, to read books twice.
Your banjo tunes excite the mice!

All is fine, until you're hurt.
Your body sounds a shrill alert.
Special cells lie dormant, sleeping,
Spring to action, almost leaping.

The stem cell team lives in your body,
Replaces parts when they are shoddy.
They rush to fix,
Come full of tricks,
They replicate, they specialize,
Build soft smooth skin, and big bright eyes.

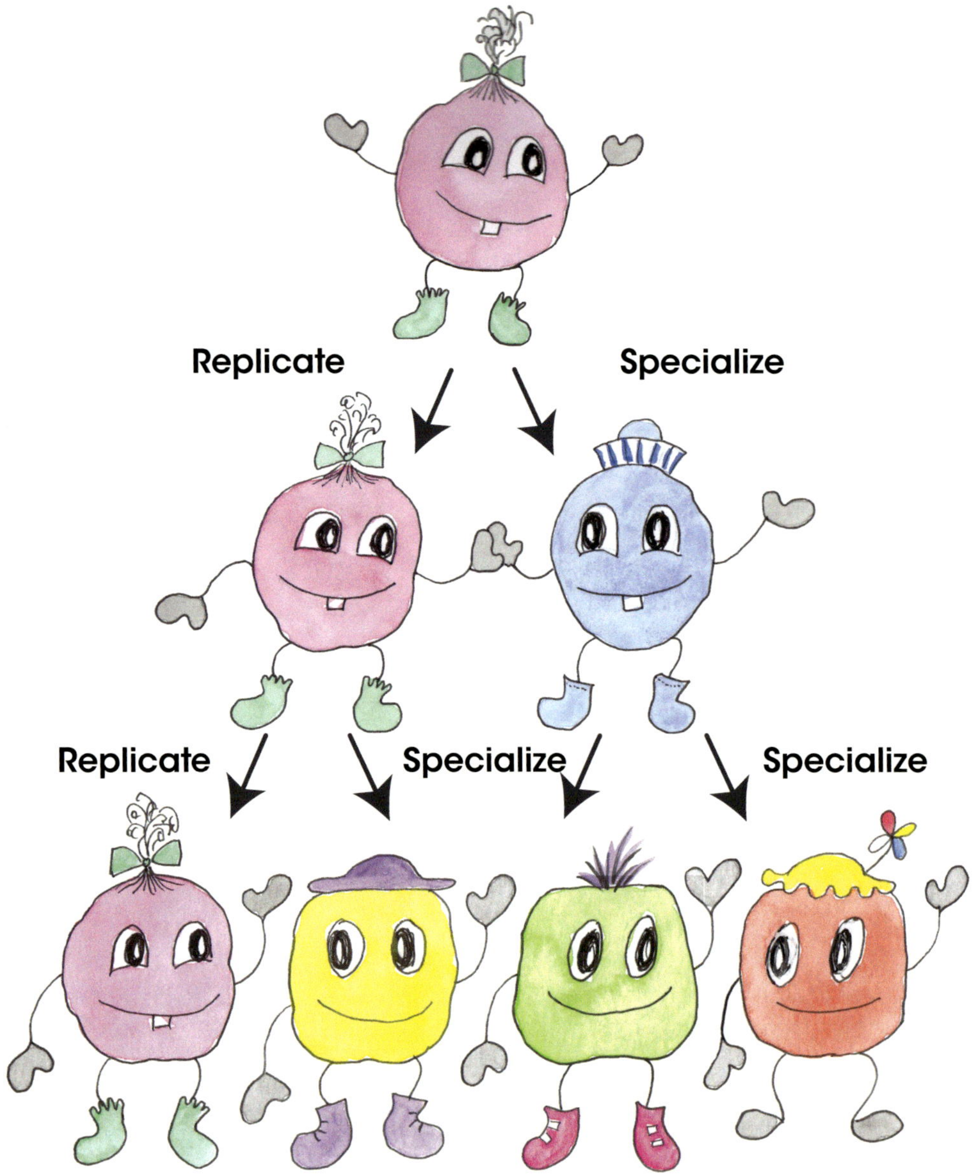

Did you know that even you
Were once a stem cell? Well, it's true!

Let's find out now, just what they do,
The stem cell team inside of you,
To come to the rescue, cells renew.

Calling once, calling twice, here come the stem cells, so precise!

While playing soccer in the rain,
You fall, your muscles ache in pain.
You groan, cry "ouch!," your outlook's bleak,
Upset you can't play at your peak.

Rousing from their deepest slumbers,
Stem cells stir, increase their numbers.
They replicate, they infiltrate,
They morph to muscle at a rapid rate.

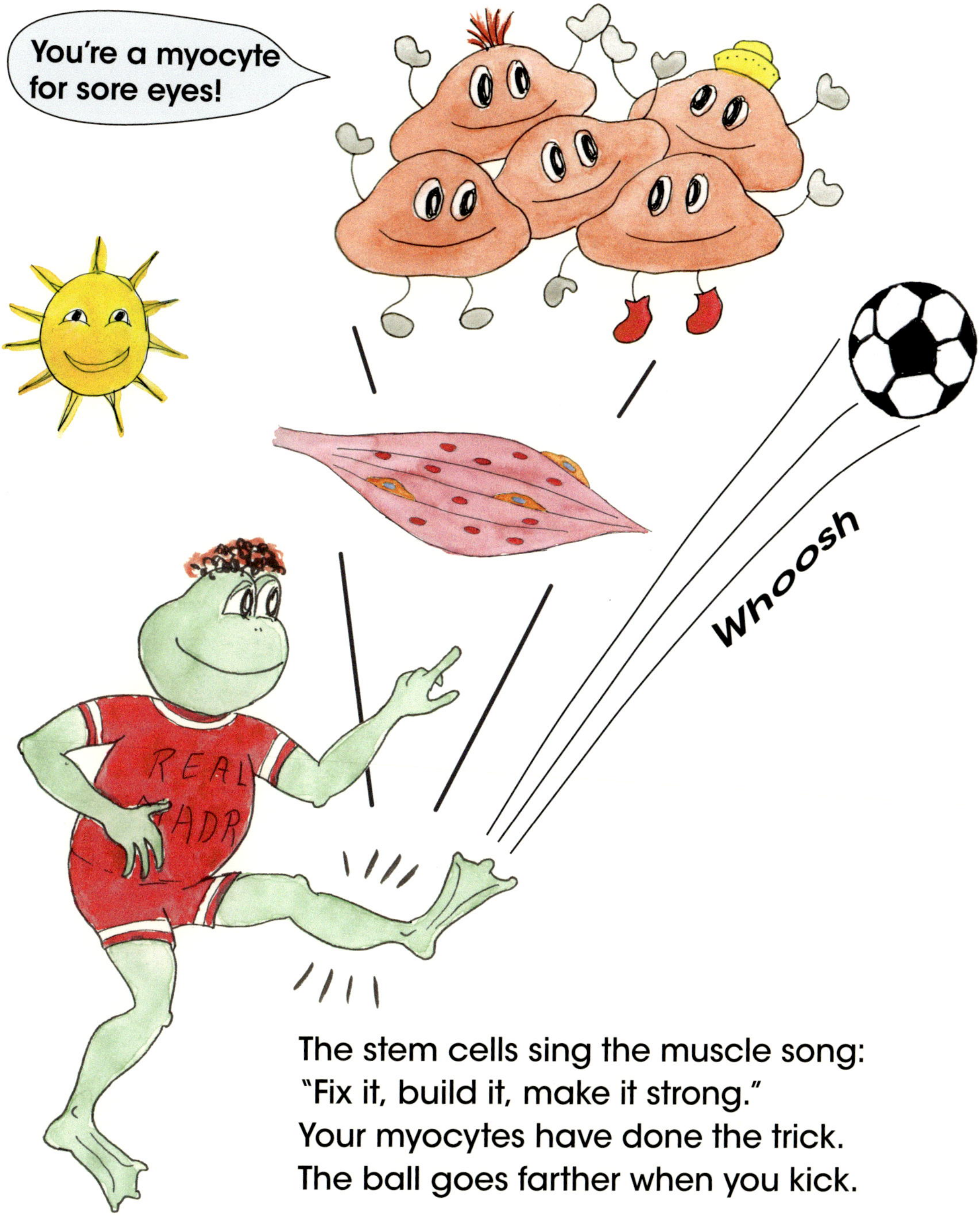

The stem cells sing the muscle song:
"Fix it, build it, make it strong."
Your myocytes have done the trick.
The ball goes farther when you kick.

Kicking once, kicking twice, being strong is paradise.

You're climbing on the jungle gym,
You slip, you fall, you break a limb.
Your broken bone sounds an alarm:
"Come stem cells quick, save me from harm!"

These special cells live in your bone,
Recruited from their central zone—
Your broken bone becomes repaired,
Now strong, you're smiling, quite prepared.
You climb the bars with both your arms.
You're good as new, now that's the charm!

*Get the message, find the wound, bone
stem cells remain attuned!*

Lying happily on the beach,
Reading, sunscreen out of reach,
You fall asleep, forget to turn.
Now you will get a reddish burn.

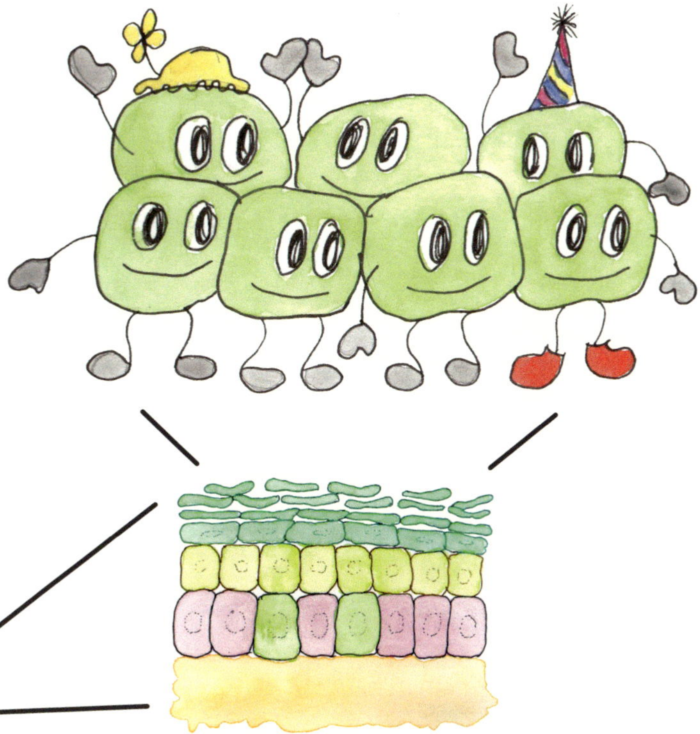

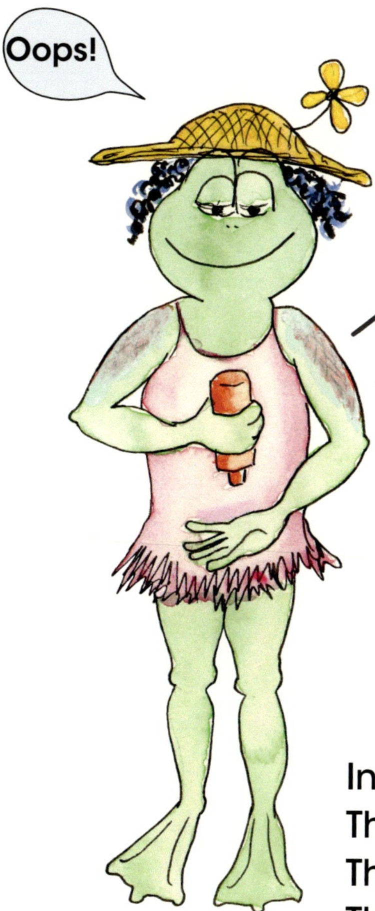

In come your stem cells from your skin.
They stop their naps, they must begin,
They sense the signals from the sore.
They multiply, make many more!
Your lovely skin they will restore.

Squeezing in, making space, skin stem cells find the perfect place.

You trip, you fall, you scrape your knee,
The blood flows red, a sight to see!
You need more blood to feed your heart,
Your brain needs blood to keep you smart.

You've got it now, how you'll get better,
The stem cells in your blood will enter.
These clever cells are multipotent.
They decide, in just one moment
What shape to take, what type to be.
They've got lots of options, as you'll see.

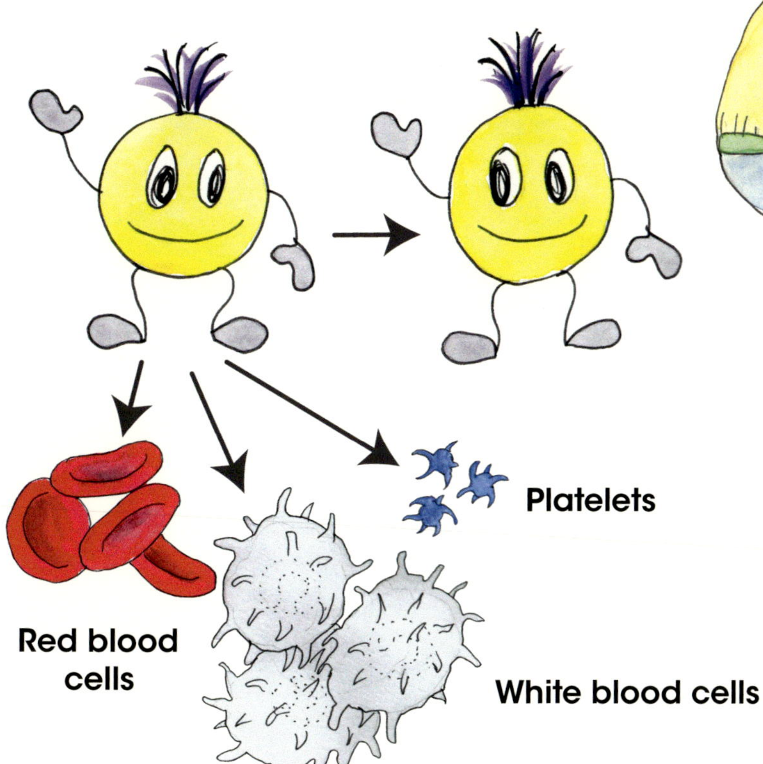

They divide in two, one will remain—
A blood stem cell you can retain.
If you need air, red cells arise.
If germs attack, white cells surprise!
To stop blood flow, you need the platelets.
To do this job, they are your favorites.

The stem cell team is always there
Your damaged tissues to repair.
But stem cells falter as we age,
The rescue team cannot engage.

Bewildered stem cells lose their way.
Old stem cells find it hard to play.

If Grandma's vision's getting blurry,
She could need stem cells in a hurry.
Her own are old and somewhat sick,
So she may need a transplant quick.
But transplants pose another issue:
Her body rejects foreign tissue.

To make fresh stem cells for her eyes,
We scientists have a big surprise.

From Grandma's blood, we get blood stem cells.
In wells with gels and chemicals,
We treat her cells, give new instructions,
So those cells can have new functions!

Now Grandma's cells have many options,
Feeding them with strong concoctions.
We tweak them, change them, and "my oh my,"
Now they're stem cells for the eye.
Put in her retina where they should be,
We hope that Grandma starts to see.

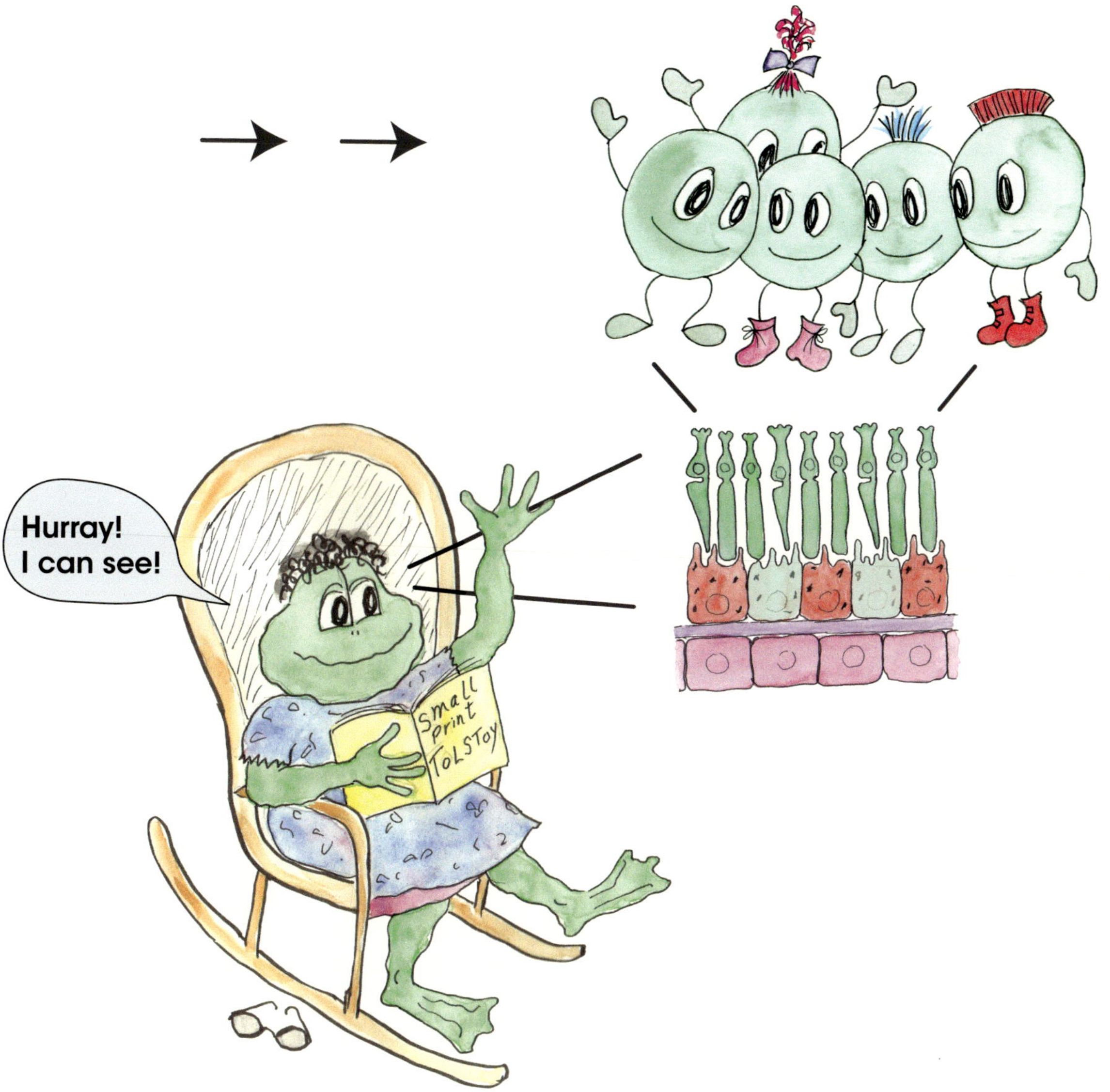

Your Grandpa's knees hurt, such a pain!
But he gets around with his trusty cane.
He needs surgery, knee replacement,
To reach his projects in the basement.

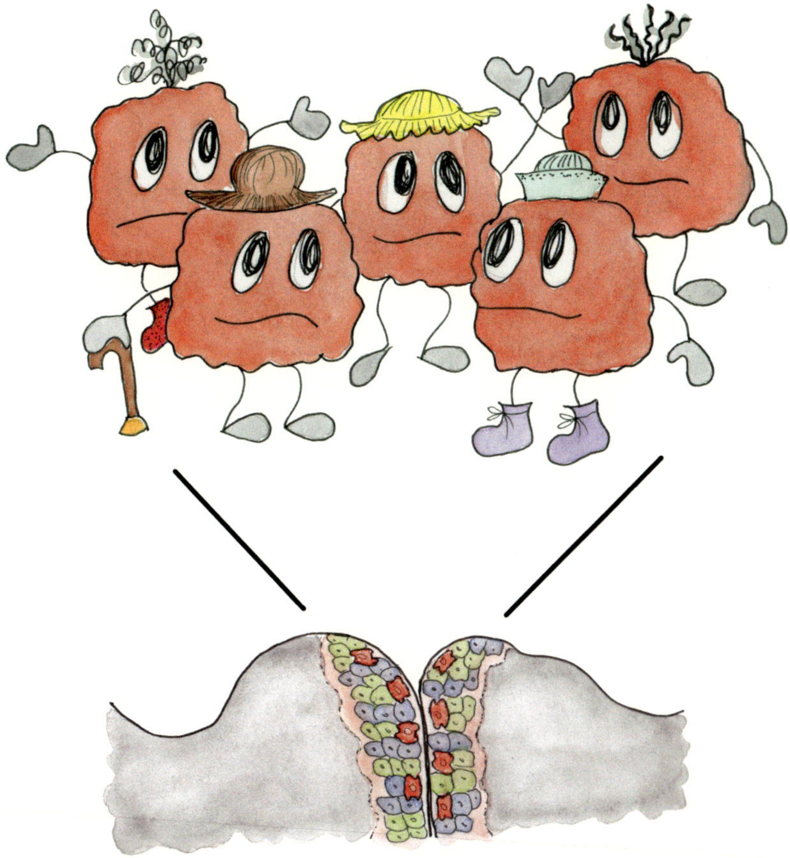

Aged stem cells cannot mend the joint,
Can we restore them? We're near that point!

Come rescue, stem cells to the fore, we need you now, to come restore!

You need your ears, so you can hear
What friends are saying or when cars are near,
And hooting owls and purring cats,
And violins and flying bats!

If Grandpa's ears cease to hear
The sounds around him won't be clear.
For him to hear what you will say,
He'll need ear stem cells right away.
To scientists, it's nearly clear
How to boost the stem cells in the ear.

Pinging once, pinging twice, can stem cells become a hearing device?

And what about our special brains?
Can stem cells help us recall names?
Cells have been found, but what to do?
We'll prod them, get them to renew.

When we grow old, we need these cells
To make good copies of themselves,
Replace the damage that ensues
As we age and functions lose.

Some answers are not here today.
We have to find out what we may.
How can we change the stem cell state?
Make aged cells rejuvenate?

To work out what makes stem cells tick,
Scientists use every trick!
Find those stem cells, make them grow,
Label them and make them glow!
Track them, treat them, 'til we know
How to make them go, Go, GO!

Make stem cells work where they are needed.
That's the goal, when we've succeeded.
Stem cells are a great solution
To fix hurt parts, find resolution.
Those special cells so very small,
Come to the rescue to help us all!

Glossary

TYPES OF STEM CELLS

There are three kinds of stem cells: embryonic stem (ES) cells, induced pluripotent stem cells (iPSCs), and tissue-specific stem cells. ES cells and iPSCs are capable of replacing, regenerating, and restoring the cell types in your body.

Embryonic stem (ES) cells are cells derived from embryos. Although invaluable for laboratory studies and used in some clinical settings, their need has been largely overcome by the relatively recent development of induced pluripotent stem cells (iPSCs) that have similar properties to ES cells, but do not require an embryo.

Induced pluripotent stem cells: From specialized cells obtained from a person's blood, urine, or skin, it is now possible to create pluripotent stem cells (induced pluripotent stem cells, iPSCs) in the laboratory that are similar in their properties to embryonic stem cells. ES cells and iPSCs can be coaxed to become essentially any cell type in the body (e.g., the eye, pp. 21 and 22). iPSCs can specialize to become essentially any desired cell type.

Tissue-specific stem cells: In many tissues of your body, there are stem cells that are dedicated to the repair of that tissue. They are generally present in a quiescent state (dormant) until there is injury, whereupon they spring into action. They divide (replicate to make copies of themselves), differentiate (specialize to participate in tissue function), and regenerate the tissue (e.g., p. 10). When you are young, your stem cells are very effective in repairing damage to muscle, bone, skin, and blood, as well as many other tissues (pp. 12–19). With aging, the stem cells become less potent. Scientists are actively seeking ways to enhance and rejuvenate the function of all stem cells, including aged stem cells (e.g., cartilage, pp. 24 and 25; ear stem cells, pp. 26 and 27; brain, pp. 28 and 29).

TERMS AND TREATMENTS

Blood stem cells (hematopoietic stem cells) are multipotent: They are capable of making copies of themselves and all the specialized cell types in the blood including ***red blood cells***, which carry oxygen to the tissues, ***white blood cells***, which fight infectious agents like bacteria and viruses, and ***platelets***, which promote clotting after injury.

Brain: Scientists have identified the stem cells responsible for learning and memory in the brain, but to date drugs that can target them and augment their function have yet to be identified. Wouldn't it be exciting if a drug or stem cell transplant could help an elderly person recall names?

Cartilage is the strong, flexible connective tissue that protects your joints and bone. Cartilage is often lost after a sports injury or with aging. Stem cell therapies could potentially overcome the need for knee and hip replacements.

Central zone of bone is where the stem cells reside.

Dormant: Quiescent or sleeping—inactive. Activation usually happens in response to damage.

Hearing: Stem cells have been identified in the ear, and drugs are being sought and tested that can stimulate them to restore hearing. This could potentially overcome the need for hearing aids.

Infiltrate: Penetrate, insert into the tissue.

Macular degeneration can occur with aging and leads to vision impairment due to progressive atrophy of the retinal pigment epithelial (RPE) layer. The example here of an experimental stem cell therapy to treat macular degeneration uses iPSCs, but ES cells are also being used to replace the RPE and help restore sight.

Morph to muscle: To acquire the shape and characteristics of muscle tissue.

Myocytes: The specialized progeny of muscle stem cells. Drugs that stimulate muscle repair are actively being developed.

Rejuvenate: Become more like their younger versions.

Replicate: Make copies of themselves.

Retina is a multilayered structure in the inner lining of the eye, where the photoreceptors reside that detect light and convert photons to electrical signals. The RPE (retinal pigment epithelium) cell layer is next to the photoreceptors and helps sustain them and maintain vision.

Tissue Illustration Sources

We wish to thank the many scientists whose work enabled and inspired our book.

General Reference

Blau HM, Daley GQ. 2019. Stem cells in the treatment of disease. *N Engl J Med* **380:** 1748–1760. Our muscle tissue after Fig. 3 and blood tissue after Fig. 4.

Muscle Stem Cells

Fuchs E, Blau HM. 2020. Tissue stem cells: architects of their niches. *Cell Stem Cell* **27:** 532–556. Our muscle tissue after Fig. 6B.

Ho ATV, Palla AR, Blake MR, Yucel ND, Wang YX, Magnusson KEG, Holbrook CA, Kraft PE, Delp SL, Blau HM. 2017. Prostaglandin E2 is essential for efficacious skeletal muscle stem-cell function, augmenting regeneration and strength. *Proc Natl Acad Sci* **114:** 6675-6684. Our muscle tissue after Fig. 6.

Loreti M, Sacco A. 2022. The jam session between muscle stem cells and the extracellular matrix in the tissue microenvironment. *NPJ Regenerative Medicine* **7:** 16. Our muscle tissue after Fig. 1.

Bone

Duda GN, Geissler S, Checa S, Tsitsilonis S, Petersen A, Schmidt-Bleek K. 2023. The decisive early phase of bone regeneration. *Nat Rev Rheumatol* **19:** 78–95. Our bone tissue after Fig. 2.

Skin

Fuchs E, Blau HM. 2020. Tissue stem cells: architects of their niches. *Cell Stem Cell* **27:** 532–556. Our skin tissue after Fig. 1A.

Solanas G, Benitah SA. 2013. Regenerating the skin: a task for the heterogeneous stem cell pool and surrounding niche. *Nat Rev Mol Cell Biol* **14:** 737–748. Our skin tissue adapted from Fig. 2, with permission from Springer Nature.

Blood

Choudhary OP, Sarkar R, Priyanka, Chethan GE, Doley PJ, Kalita PC, Kalita A. 2021. Preparation of blood samples for electron microscopy: the standard protocol. *Ann Med Surg* **70:** 1–4. Our blood tissue after Figs. 1 and 3.

Comazzetto S, Shen B, Morrison SJ. 2021. Niches that regulate stem cells and hematopoiesis in adult bone marrow. *Dev Cell* **56:** 1848–1860. Our blood tissue after Fig. 1.

Age-Related Macular Degeneration of Retina of Eye

Fain GL. 2020. Photoreception, Chap. 9. In *Sensory transduction*, 2nd ed. Oxford University Press, New York. (Illustrations by Margery J. Fain.) Our eye tissue after Figs. 9-13 and 9-25.

Stern JH, Tian Y, Funderburgh J, Pellegrini G, Zhang K, Goldberg JL, Ali RR, Young M, Xie Y, Temple S. 2018. Regenerating eye tissues to preserve and restore vision. *Cell Stem Cell* **23:** 453. Our eye tissue after Fig. 1.

Cartilage

Murphy MP, Koepke LS, Lopez MT, Tong X, Ambrosi TH, Gulati GS, Marecic O, Wang Y, Ransom RC, Hoover MY, et al. 2020. Articular cartilage regeneration by activated skeletal stem cells. *Nat Med* **26:** 1583–1592. Our cartilage tissue after Fig. 1.

Smeriglio P, Grandi FC, Davala S, Masarapu V, Indelli PF, Goodman SB, Bhutani N. 2020. Inhibition of TET1 prevents the development of osteoarthritis and reveals the 5hmC landscape that orchestrates pathogenesis. *Sci Transl Med* **12:** 1–12. Our cartilage tissue after Fig. 2.

Hearing

Benkafadar N, Sato MP, Ling AH, Janesick A, Scheibinger M, Jan TA, Heller S. 2024. An essential signaling cascade for avian auditory hair cell regeneration. *Dev Cell* **59:** 280–291.e5. Our hearing tissue after Fig. 1.

Fain GL 2020. Hearing and hair cells, Chap. 6. In *Sensory transduction*, 2nd ed. Oxford University Press, New York. (Illustrations by Margery J. Fain.) Our hearing tissue after Fig. 6-17.

Neural Stem Cells in Brain

Conover JC, Shook BA. 2011. Aging of the subventricular zone neural stem cell niche. *Aging Dis* **2:** 49–63. Our brain tissue adapted from Fig. 1, under CC-BY-NC-ND 4.0 license.

Gonçalves JT, Schafer ST, Gage FH. 2016. Adult neurogenesis in the hippocampus: from stem cells to behavior. *Cell* **167:** 897–914. Our brain tissue after Fig. 2.